Earth in Balance

EARTH DAY 2024

Sophia Reynolds

All rights reserved. No part of this publication may be reproduced, distributed, or transmitted in any form or by any means, including photocopying, recording, or other electronic or mechanical methods, without the prior written permission of the publisher, except in the case of brief quotations embodied in critical reviews and certain other noncommercial uses permitted by copyright law.

TABLE OF CONTENTS

Introduction 1
Setting the scene for Earth Day 2024.. 1
Importance of environmental awareness and action 3

The State of the Planet.. 5
Transportation Options 5
Climate change impacts 7
Biodiversity loss 9
Pollution and waste.. 12

Global Efforts.. 15
International agreements and initiatives................................. 15
Role of governments, organizations, and activists................... 19
Success stories and ongoing challenges 22

Local Initiatives... 25
Community-based projects... 25
Grassroots movements... 28
Individuals making differences in their communities 31

Technology and Innovation.. 33
Advances in renewable energy... 33
Sustainable practices in industry and agriculture.................... 37
Tech solutions for environmental monitoring......................... 41

Youth Activism 44
Role of young people in driving environmental change........... 44
Student-led movements and campaigns 47
Youth voices and perspectives on the future of the planet 50

Conservation and Restoration 52

Efforts to protect ecosystems and endangered species............. 52
Restoration projects 55
Importance of preserving natural spaces 58
Sustainable Living 60
Tips for reducing personal environmental impact.................... 60
Eco-friendly habits and practices 63
Green consumerism and ethical choices 66
The Future of Earth Day 69
Reflections on the evolution of Earth Day 69
Vision for the future of environmental activism 72
Call to action for readers 75
Hope for the planet and its inhabitants 77

Voices for the Earth

Welcome to "Earth in Balance: Earth Day 2024." In the midst of unprecedented environmental challenges, we stand at a critical juncture in human history. The urgency of addressing climate change, biodiversity loss, pollution, and habitat destruction has never been more evident. Yet, alongside these challenges lies a profound opportunity for positive change and transformation.

As we mark Earth Day 2024, we are reminded of the significance of this annual observance - a time to reflect on our relationship with the planet, celebrate its beauty and diversity, and renew our commitment to environmental stewardship. This book serves as a comprehensive exploration of the state of our planet, the interconnectedness of environmental issues, and the pathways to a more sustainable and resilient future.

Drawing on the latest scientific research, expert insights, and real-world examples, "Earth in Balance" delves into the pressing environmental challenges facing our world today. From the impacts of climate change on communities and ecosystems to the loss of biodiversity and the need for sustainable development, each chapter provides a thorough analysis of key issues and trends shaping the environmental landscape.

But this book is not just about the problems we face - it is about the solutions and opportunities that lie ahead. Through a lens of hope, optimism, and empowerment, we explore innovative strategies, grassroots initiatives, and collective actions that are driving positive change and shaping a more sustainable world.

From renewable energy and regenerative agriculture to youth activism and community resilience, "Earth in Balance" showcases the diverse array of efforts and initiatives contributing to environmental conservation and sustainability. By highlighting inspiring stories, practical tips, and actionable insights, we aim to inform, inspire, and empower readers to become agents of change in their own communities and beyond.

As we embark on this journey together, let us recommit ourselves to the vision of a world where people and the planet thrive in harmony. Let us embrace the challenge and opportunity of our time with courage, determination, and optimism. Together, we can work towards a future where Earth is truly in balance - a future of hope, resilience, and sustainability for generations to come.

Welcome to "Earth in Balance: Earth Day 2024." Let the journey begin.

> *Chapter One*

Introduction

Setting the scene for Earth Day 2024

As dawn breaks on April 22, 2024, the world awakens to a planet at a crossroads. The sun's gentle rays illuminate landscapes scarred by the relentless march of industrialization and human activity. From bustling cities to remote wilderness areas, the signs of environmental degradation are unmistakable.

In urban centers, skyscrapers pierce the skyline, their glass facades reflecting the hazy light of dawn. Below, the streets teem with activity as millions of people begin their daily routines. Traffic jams clog the roads, filling the air with exhaust fumes and the incessant honking of horns. Despite efforts to curb pollution, the air quality remains poor, casting a gray pall over the cityscape.

Meanwhile, in rural regions, the landscape tells a different story. Fields once lush with greenery now lie barren and parched, victims of prolonged droughts exacerbated by climate change. Rivers and lakes, once lifelines for communities, have dwindled to mere shadows of their former selves, their waters polluted and unfit for consumption. Forests, once teeming with life, have been reduced to fragmented patches of wilderness, their inhabitants struggling to survive in a world increasingly hostile to their existence.

But amidst the bleakness, there are glimmers of hope. Across the globe, individuals and communities have rallied together in defense of the planet. From grassroots activists to high-profile celebrities, voices calling for environmental stewardship grow louder by the day. Innovative technologies and sustainable practices offer glimpses of a greener future, where humanity lives in harmony with the natural world rather than exploiting it for short-term gain.

On this Earth Day, the world stands at a critical juncture. The challenges we face are daunting, but the resolve to overcome them is stronger than ever. As the sun rises higher in the sky, casting its warm glow over the Earth, there is a sense of urgency in the air. For in the hands of each individual lies the power to shape the destiny of our planet. As the day unfolds, the stage is set for a global movement of unprecedented scale and significance—a movement determined to chart a course towards a more sustainable and equitable future for all.

Importance of environmental awareness and action

In an era defined by rapid industrialization, technological advancement, and exponential population growth, the health of our planet hangs in the balance. Environmental awareness serves as a beacon, illuminating the interconnectedness between human activities and the delicate ecosystems that sustain life on Earth. At its core, environmental awareness is about recognizing the profound impact of our actions on the natural world and acknowledging our responsibility to preserve and protect it for future generations.

One of the primary reasons why environmental awareness is crucial lies in its role as a catalyst for informed decision-making. By understanding the intricate web of ecological relationships and the consequences of our choices, individuals, communities, and policymakers can make more sustainable and ethical decisions. Whether it's opting for renewable energy sources, reducing consumption and waste, or advocating for conservation measures, informed actions driven by environmental awareness have the power to mitigate environmental degradation and foster resilience in the face of global challenges.

Moreover, environmental awareness fosters a deeper sense of connection and stewardship towards the natural world. By recognizing the inherent value of biodiversity and the intrinsic worth of ecosystems beyond their instrumental utility to humans, individuals are inspired to act as custodians of the Earth rather than mere exploiters of its resources. This shift in perspective is essential for cultivating a culture of sustainability and fostering a harmonious relationship between humanity and nature.

However, environmental awareness alone is not sufficient to address the myriad environmental crises we face. Action is equally imperative, serving as the tangible manifestation of our commitment to environmental stewardship. From grassroots initiatives to international treaties, from individual lifestyle changes to corporate sustainability practices, action takes many forms and operates at multiple scales. What unites these diverse efforts is a shared goal: to effect positive change and safeguard the integrity of our planet.

The importance of environmental action extends beyond the preservation of biodiversity and ecosystem services; it also encompasses social and economic dimensions. Environmental degradation disproportionately impacts marginalized communities, exacerbating inequalities and perpetuating cycles of poverty and vulnerability. By prioritizing environmental action, we not only safeguard the natural world but also promote social justice, equity, and human well-being.

Furthermore, environmental action is a testament to the power of collective agency and solidarity. In a world fraught with division and discord, the fight for environmental sustainability unites people across geographical, cultural, and ideological boundaries. It transcends individual interests and short-term gains, calling upon us to work together towards a common purpose: the protection of our shared home, Earth.

In conclusion, environmental awareness and action are inseparable components of a holistic approach to environmental conservation and sustainability. Together, they form the cornerstone of a global movement striving to safeguard the future of our planet and ensure a prosperous and thriving existence for all life forms. As stewards of this magnificent blue planet, it is incumbent upon us to heed the call of environmental awareness and translate it into meaningful action. For the fate of our world and the generations to come hangs in the balance, and the time to act is now.

Chapter Two

The State of the Planet

Transportation Options

The 21st century presents humanity with an array of unprecedented environmental challenges, stemming from a combination of population growth, industrialization, urbanization, and unsustainable consumption patterns. These challenges threaten the very foundation of life on Earth, jeopardizing ecosystems, biodiversity, and the well-being of present and future generations.

1. **Climate Change**: Perhaps the most pressing environmental challenge of our time, climate change is driven primarily by the emission of greenhouse gases from human activities such as burning fossil fuels, deforestation, and industrial processes. The consequences of climate change are far-reaching and severe, including rising temperatures, extreme weather events, sea-level rise, disruptions to ecosystems, and threats to food security and water resources.
2. **Biodiversity Loss**: The rapid loss of biodiversity is another critical environmental challenge, driven by habitat destruction, pollution, overexploitation of natural resources, and invasive species. This loss of biodiversity not only diminishes the resilience of ecosystems but also undermines essential ecosystem services such as pollination, nutrient cycling, and water purification, upon which human societies depend.
3. **Pollution**: Pollution in its various forms—air, water, soil, and noise—poses significant threats to human health, ecosystems, and biodiversity. Air pollution from vehicle emissions, industrial processes, and agricultural practices contributes to respiratory diseases, cardiovascular problems, and premature mortality. Water pollution from untreated

sewage, industrial discharge, and agricultural runoff contaminates freshwater sources, compromising drinking water quality and aquatic ecosystems. Soil pollution, often resulting from industrial activities and improper waste disposal, degrades soil fertility and undermines agricultural productivity.
4. **Resource Depletion**: Humanity's ever-increasing demand for natural resources, coupled with unsustainable extraction and consumption patterns, is depleting finite resources such as freshwater, minerals, and forests at an alarming rate. This depletion not only threatens ecosystems and biodiversity but also exacerbates social inequalities and conflicts over scarce resources.
5. **Loss of Ecosystem Services**: Ecosystems provide a wide range of essential services that support life on Earth, including provisioning services (e.g., food, water, timber), regulating services (e.g., climate regulation, flood control), cultural services (e.g., recreation, spiritual value), and supporting services (e.g., nutrient cycling, soil formation). However, human activities are degrading and destroying ecosystems at an unprecedented rate, compromising their ability to provide these vital services.
6. **Ocean Degradation**: The world's oceans face numerous threats, including overfishing, habitat destruction, pollution, ocean acidification, and climate change. These threats pose significant risks to marine biodiversity, fisheries, coastal communities, and the health of the entire planet, given the oceans' critical role in regulating the climate and supporting life.
7. **Urbanization and Land Use Change**: The rapid expansion of urban areas and the conversion of natural habitats for agricultural and urban development are altering landscapes and fragmenting ecosystems, leading to loss of biodiversity, habitat degradation, and increased vulnerability to natural disasters.
8. **Waste Management**: The generation of vast quantities of solid waste, including plastics, electronic waste, and hazardous materials, poses significant environmental and health risks. Inadequate waste management practices, including improper disposal and lack of recycling infrastructure, contribute to pollution of land, water, and air, further exacerbating environmental degradation.

Climate change impacts

Climate change impacts are wide-ranging and profound, affecting ecosystems, societies, economies, and human well-being in diverse ways. Here's a detailed exploration of some of the key impacts:

1. **Rising Temperatures**: One of the most obvious and well-documented impacts of climate change is the steady rise in global temperatures. This trend leads to more frequent and intense heatwaves, which can have serious implications for human health, particularly among vulnerable populations such as the elderly and those with pre-existing health conditions. High temperatures also exacerbate water scarcity, increase energy demand for cooling, and disrupt agricultural productivity.
2. **Extreme Weather Events**: Climate change is intensifying many types of extreme weather events, including hurricanes, cyclones, floods, droughts, and wildfires. These events can result in devastating loss of life, displacement of populations, destruction of infrastructure, and disruption of essential services such as food and water supply. The economic costs associated with extreme weather events are staggering, imposing a heavy burden on communities, governments, and economies worldwide.
3. **Sea-Level Rise**: As global temperatures rise, glaciers and ice caps melt, contributing to sea-level rise. This phenomenon poses a grave threat to coastal communities, infrastructure, and ecosystems. Rising sea levels increase the risk of coastal flooding, erosion, and saltwater intrusion into freshwater sources, jeopardizing the livelihoods of millions of people who depend on coastal ecosystems for their sustenance and well-being.
4. **Ocean Acidification**: The absorption of excess carbon dioxide by the oceans is causing ocean acidification, which has profound implications for marine ecosystems. Acidification can disrupt the growth and survival of calcifying organisms such as corals, shellfish, and plankton, which form the foundation of marine food webs. This, in turn, can ripple through marine ecosystems, impacting fish stocks, biodiversity, and the livelihoods of coastal communities dependent on fisheries.

5. **Shifts in Ecosystems and Biodiversity**: Climate change is driving shifts in the distribution and composition of ecosystems worldwide. Species are migrating towards cooler regions or higher altitudes in response to changing climatic conditions, leading to alterations in species interactions, food webs, and ecosystem functioning. Some species may face increased risk of extinction if they are unable to adapt or migrate to suitable habitats.
6. **Water Scarcity and Changes in Precipitation Patterns**: Changes in precipitation patterns, including altered timing, intensity, and distribution of rainfall, are exacerbating water scarcity in many regions. This has significant implications for agriculture, water supply, hydropower generation, and ecosystem health. Competition for dwindling water resources can exacerbate tensions between different sectors and communities, leading to conflicts over access to water.
7. **Food Insecurity and Agricultural Disruption**: Climate change poses a threat to global food security by disrupting agricultural systems and reducing crop yields. Extreme weather events, changing precipitation patterns, and temperature extremes can damage crops, reduce livestock productivity, and degrade soil fertility. Smallholder farmers, particularly those in low-income countries with limited adaptive capacity, are disproportionately affected by these impacts, exacerbating poverty and malnutrition.
8. **Health Impacts**: Climate change affects human health through various pathways, including heat stress, air pollution, vector-borne diseases, and food insecurity. Rising temperatures can exacerbate heat-related illnesses and lead to increased mortality rates during heatwaves. Poor air quality, exacerbated by higher temperatures and increased wildfires, can worsen respiratory and cardiovascular diseases. Changes in precipitation patterns can alter the distribution of disease vectors such as mosquitoes, leading to the spread of diseases such as malaria, dengue fever, and Zika virus.

Biodiversity loss

Biodiversity loss is a critical environmental issue that poses significant threats to ecosystems, species, and the well-being of human societies. Here's a detailed exploration of the causes, consequences, and implications of biodiversity loss:

1. **Causes of Biodiversity Loss**:

    A. **Habitat Destruction**: The primary driver of biodiversity loss is habitat destruction and fragmentation, resulting from human activities such as deforestation, urbanization, agriculture, and infrastructure development. These activities destroy and degrade natural habitats, leaving species with limited space and resources to survive.

    B. **Climate Change**: Climate change is exacerbating biodiversity loss by altering ecosystems, disrupting species distributions, and increasing the frequency and intensity of extreme weather events. Species that are unable to adapt or migrate to suitable habitats face increased risk of extinction.

    C. **Overexploitation of Natural Resources**: Unsustainable exploitation of natural resources, including overfishing, illegal logging, and poaching, threatens the survival of many species. Overharvesting of wild populations can lead to population declines and ecosystem imbalances.

    D. **Pollution**: Pollution in its various forms, including air pollution, water pollution, and soil contamination, poses significant threats to biodiversity. Chemical pollutants, plastic waste, and oil spills can harm wildlife, disrupt ecosystems, and degrade habitat quality.

    E. **Invasive Species**: Introduction of non-native species into new environments can have detrimental effects on native biodiversity. Invasive species can outcompete native species for resources, prey upon them, or introduce diseases, leading to population declines and ecosystem disruptions.

2. **Consequences of Biodiversity Loss:**

    A. **Ecosystem Instability**: Biodiversity loss undermines the resilience of ecosystems, making them more vulnerable to environmental disturbances such as climate change, disease outbreaks, and invasive species. Loss of key species can disrupt ecosystem functions and processes, leading to reduced productivity and stability.

    B. **Impacts on Ecosystem Services**: Biodiverse ecosystems provide a wide range of essential services that support human well-being, including pollination, nutrient cycling, water purification, and climate regulation. Biodiversity loss diminishes the capacity of ecosystems to provide these services, jeopardizing food security, water quality, and other critical resources.

    C. **Economic Costs**: Biodiversity loss can have significant economic costs, both directly and indirectly. Declines in wild populations can impact industries such as fisheries, forestry, and agriculture, leading to reduced yields, increased production costs, and loss of livelihoods. Additionally, the loss of ecosystem services can result in costly human interventions to mitigate the impacts.

    D. **Cultural and Aesthetic Value**: Biodiversity has intrinsic value beyond its instrumental utility to humans. It enriches our lives culturally, spiritually, and aesthetically, providing inspiration for art, literature, and recreation. Loss of biodiversity can erode cultural heritage and diminish the quality of life for future generations.

3. **Implications for Human Well-being:**

    A. **Food Security**: Biodiversity loss threatens global food security by reducing genetic diversity in crops and livestock, making agricultural systems more susceptible to pests, diseases, and environmental stresses. Diverse ecosystems also support wild foods, medicinal plants, and traditional knowledge systems that are essential for human nutrition and health.

B. **Health Impacts**: Biodiversity loss can have direct and indirect impacts on human health. Loss of biodiversity can increase the risk of zoonotic diseases, such as Ebola and COVID-19, by altering wildlife populations and their habitats. Additionally, biodiversity loss can reduce access to medicinal plants and compounds used in traditional medicine.

C. **Livelihoods and Economic Development**: Many communities depend directly on biodiversity for their livelihoods, including indigenous peoples, fisherfolk, and farmers. Loss of biodiversity can undermine these livelihoods, leading to poverty, social unrest, and forced migration. Additionally, biodiversity loss can hinder economic development by limiting opportunities for ecotourism, bioprospecting, and sustainable resource management.

Pollution and waste

Pollution and waste are pervasive environmental problems that pose significant threats to ecosystems, human health, and the well-being of communities worldwide. Here's a detailed exploration of their causes, consequences, and implications:

1. **Types of Pollution and Waste:**

A. **Air Pollution**: Air pollution refers to the presence of harmful substances, such as particulate matter, nitrogen oxides, sulfur dioxide, volatile organic compounds, and heavy metals, in the air. Sources of air pollution include industrial emissions, vehicle exhaust, agricultural activities, and wildfires.

B. **Water Pollution**: Water pollution occurs when pollutants contaminate freshwater sources such as rivers, lakes, and groundwater. Common water pollutants include pathogens, nutrients (e.g., nitrogen and phosphorus), heavy metals, pesticides, and pharmaceuticals. Sources of water pollution include industrial discharge, sewage effluent, agricultural runoff, and improper waste disposal.

C. **Soil Pollution**: Soil pollution, also known as land contamination, occurs when hazardous substances accumulate in the soil, compromising soil quality and fertility. Soil pollutants include heavy metals, pesticides, industrial chemicals, and petroleum hydrocarbons. Soil pollution can result from industrial activities, mining operations, improper waste disposal, and agricultural practices.

D. **Plastic Pollution**: Plastic pollution refers to the accumulation of plastic debris in the environment, including oceans, rivers, lakes, and terrestrial ecosystems. Single-use plastics, such as bottles, bags, and packaging, are major contributors to plastic pollution. Plastic pollution poses serious threats to wildlife, ecosystems, and human health, as plastics can persist in the environment for hundreds of years, leach harmful chemicals, and entangle or be ingested by marine animals.

E. **Noise Pollution**: Noise pollution, also known as sound pollution, refers to excessive or unwanted noise that disrupts the environment and adversely affects human health and well-being. Sources of noise pollution include transportation (e.g., traffic, aircraft, railways), industrial activities, construction sites, and recreational activities (e.g., concerts, sporting events).

2. **Consequences of Pollution and Waste**:

   A. **Human Health Impacts**: Pollution and waste pose significant risks to human health, contributing to respiratory diseases, cardiovascular problems, neurological disorders, and cancers. Air pollution, for example, is a leading cause of premature mortality worldwide, while waterborne diseases resulting from contaminated water sources cause millions of deaths annually, particularly in developing countries.

   B. **Ecosystem Degradation**: Pollution and waste degrade ecosystems, disrupt ecological processes, and diminish biodiversity. Water pollution, for instance, can lead to algal blooms, fish kills, and loss of aquatic habitats, threatening the survival of species and undermining ecosystem resilience. Plastic pollution in marine environments poses particular risks to marine life, as animals may ingest or become entangled in plastic debris, leading to injury, suffocation, or starvation.

   C. **Environmental Degradation**: Pollution and waste contribute to environmental degradation, including degradation of air quality, water quality, and soil fertility. Soil pollution, for example, can render land unsuitable for agriculture or other productive uses, leading to land degradation and loss of ecosystem services. Noise pollution disrupts wildlife behavior and communication, alters ecosystems, and can have physiological and psychological effects on humans.

D. **Social and Economic Costs**: Pollution and waste impose significant social and economic costs on communities and societies. Health care costs associated with pollution-related illnesses, cleanup and remediation costs for contaminated sites, and loss of ecosystem services all contribute to economic burdens. Additionally, pollution-related impacts on tourism, fisheries, agriculture, and other industries can result in lost revenue and livelihoods.

3. **Implications for Sustainable Development**:

    A. **Sustainable Consumption and Production**: Addressing pollution and waste requires transitioning towards sustainable consumption and production patterns that minimize resource use, reduce waste generation, and prioritize the use of environmentally friendly materials and technologies.

    B. **Circular Economy**: Embracing the principles of a circular economy, which seeks to minimize waste and maximize resource efficiency through reuse, recycling, and recovery, can help mitigate pollution and waste and promote sustainable development.

    C. **Policy and Regulation**: Effective policies and regulations are essential for addressing pollution and waste at local, national, and global levels. Measures such as emission standards, pollution taxes, waste management regulations, and bans on single-use plastics can help reduce pollution and promote responsible waste management practices.

    D. **Community Engagement and Education**: Community engagement, public awareness campaigns, and environmental education initiatives play crucial roles in raising awareness about pollution and waste issues, fostering behavior change, and empowering individuals and communities to take action.

Chapter Three

Global Efforts

International agreements and initiatives

International agreements and initiatives play a crucial role in addressing global environmental challenges by promoting cooperation, setting standards, and coordinating action among nations. Here's a detailed exploration of some key international agreements and initiatives related to environmental conservation and sustainability:

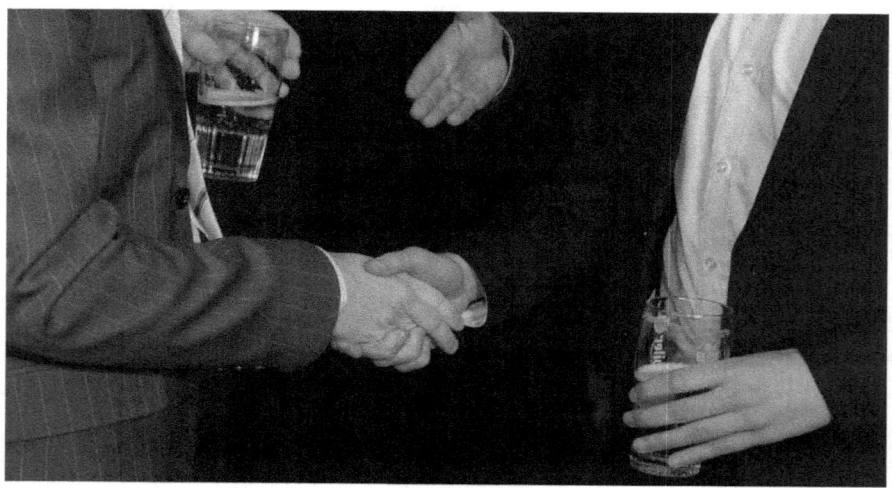

1. **United Nations Framework Convention on Climate Change (UNFCCC):**
    - Established in 1992, the UNFCCC is an international treaty aimed at combating climate change and mitigating its impacts.

- The treaty sets out general principles and goals for addressing climate change, including the stabilization of greenhouse gas concentrations in the atmosphere.
- The UNFCCC serves as the framework for subsequent agreements, including the Kyoto Protocol and the Paris Agreement, which establish binding emission reduction targets for developed and developing countries.

2. **Paris Agreement**:
   - Adopted in 2015 under the auspices of the UNFCCC, the Paris Agreement is a landmark international treaty aimed at limiting global warming to well below 2 degrees Celsius above pre-industrial levels, with efforts to limit it to 1.5 degrees Celsius.
   - The agreement commits signatory parties to nationally determined contributions (NDCs), outlining their targets for reducing greenhouse gas emissions and adapting to the impacts of climate change.
   - The Paris Agreement emphasizes the importance of international cooperation, transparency, and support for developing countries in addressing climate change.

3. **Convention on Biological Diversity (CBD)**:
   - Adopted in 1992, the CBD is an international treaty aimed at conserving biodiversity, promoting sustainable use of biological resources, and ensuring equitable sharing of benefits derived from genetic resources.
   - The CBD sets out three main objectives: conservation of biological diversity, sustainable use of its components, and fair and equitable sharing of benefits arising from genetic resources.
   - The CBD operates through various subsidiary bodies, including the Conference of the Parties (COP), which meets regularly to review progress, negotiate agreements, and adopt decisions to advance the implementation of the convention.

4. **Montreal Protocol on Substances that Deplete the Ozone Layer:**
   - Adopted in 1987, the Montreal Protocol is an international treaty aimed at phasing out the production and consumption of ozone-depleting substances (ODSs), such as chlorofluorocarbons (CFCs) and halons.
   - The protocol has been hailed as one of the most successful environmental agreements in history, leading to significant reductions in ODSs and gradual recovery of the ozone layer.
   - The Montreal Protocol demonstrates the effectiveness of international cooperation in addressing global environmental challenges and serves as a model for other multilateral agreements.

5. **Sustainable Development Goals (SDGs):**
   - Adopted by the United Nations in 2015, the SDGs are a set of 17 interconnected goals aimed at addressing global challenges, including poverty, inequality, climate change, environmental degradation, and sustainable development.
   - The SDGs provide a comprehensive framework for action, covering a wide range of social, economic, and environmental issues.
   - The SDGs emphasize the importance of partnerships and collaboration among governments, businesses, civil society, and other stakeholders in achieving sustainable development.

6. **Global Environmental Facility (GEF):**
   - Established in 1991, the GEF is a multilateral financial mechanism that provides grants and concessional funding to support projects and programs aimed at addressing global environmental challenges.
   - The GEF funds projects in areas such as biodiversity conservation, climate change mitigation and adaptation, land degradation, and sustainable forest management.
   - The GEF operates as a partnership between governments, international organizations, civil society, and the private sector, with the aim of

mobilizing resources and catalyzing action to achieve global environmental objectives.

These international agreements and initiatives demonstrate the importance of collective action and cooperation in addressing global environmental challenges. While challenges remain in implementing and enforcing these agreements, they provide important frameworks and mechanisms for advancing environmental conservation and sustainability on a global scale.

Role of governments, organizations, and activists

The role of governments, organizations, and activists is instrumental in addressing global environmental challenges, promoting sustainability, and fostering positive change. Here's a detailed exploration of their respective roles:

1. **Governments**:

   A. **Policy and Regulation**: Governments play a central role in formulating and implementing environmental policies and regulations to address pressing environmental challenges. These policies may include measures to reduce greenhouse gas emissions, protect biodiversity, promote sustainable land use and resource management, and regulate pollution and waste disposal.

   B. **International Cooperation**: Governments engage in international negotiations and agreements to address transboundary environmental issues and promote global cooperation on environmental conservation and sustainability. This includes participation in international treaties such as the Paris Agreement, the Convention on Biological Diversity, and the Montreal Protocol.

   C. **Resource Allocation**: Governments allocate financial resources, personnel, and infrastructure to support environmental protection and conservation efforts. This may involve funding for research and development of sustainable technologies, investments in renewable energy infrastructure, support for conservation initiatives, and funding for environmental education and outreach programs.

   D. **Enforcement and Compliance**: Governments are responsible for enforcing environmental laws and regulations, monitoring compliance with environmental standards, and imposing penalties on violators. Effective enforcement mechanisms help deter illegal activities and ensure accountability among individuals, businesses, and industries.

E. **Public Education and Awareness**: Governments play a key role in raising public awareness about environmental issues and promoting environmental literacy through educational programs, public campaigns, and outreach initiatives. By fostering environmental consciousness and empowering citizens to take action, governments can mobilize grassroots support for environmental conservation efforts.

2. **Non-Governmental Organizations (NGOs)**:

    A. **Advocacy and Lobbying**: NGOs advocate for environmental protection and sustainability by raising awareness about pressing environmental issues, promoting policy reforms, and lobbying governments and international institutions to take action. NGOs often serve as watchdogs, monitoring environmental developments, and holding governments and corporations accountable for their environmental impacts.

    B. **Research and Analysis**: NGOs conduct research, gather data, and provide scientific expertise on environmental issues, helping to inform policy decisions, shape public discourse, and identify solutions to environmental challenges. By generating evidence-based research, NGOs contribute to the development of effective environmental policies and strategies.

    C. **Community Engagement and Capacity Building**: NGOs work closely with local communities, indigenous peoples, and marginalized groups to empower them to participate in environmental decision-making, build their capacity to address environmental challenges, and advocate for their rights. By fostering community ownership and participation, NGOs promote sustainable development and social equity.

    D. **Direct Action and Conservation Efforts**: Many NGOs engage in direct action and conservation efforts, such as habitat restoration, wildlife protection, and environmental cleanup initiatives. These on-the-ground activities help to preserve biodiversity, restore ecosystems, and mitigate the impacts of environmental degradation.

3. **Activists:**

    A. **Awareness-Raising and Mobilization**: Environmental activists play a crucial role in raising public awareness about environmental issues, mobilizing grassroots support, and galvanizing public opinion to demand action from governments, businesses, and other stakeholders. Through protests, demonstrations, petitions, and social media campaigns, activists amplify their voices and advocate for change.

    B. **Pressure and Accountability**: Activists hold governments, corporations, and other powerful actors accountable for their environmental impacts and policies. By shining a spotlight on environmental injustices, human rights violations, and unsustainable practices, activists exert pressure on decision-makers to adopt more environmentally responsible policies and practices.

    C. **Innovation and Alternative Solutions**: Environmental activists often champion innovative solutions and alternative approaches to address environmental challenges, such as renewable energy technologies, sustainable agriculture practices, and circular economy initiatives. By promoting sustainable alternatives and challenging the status quo, activists drive innovation and change.

    D. **Coalition Building and Solidarity**: Activists collaborate with other social movements, grassroots organizations, and community groups to build coalitions, share resources, and amplify their collective impact. By forging alliances and fostering solidarity across diverse constituencies, activists strengthen their advocacy efforts and advance common goals for environmental justice and sustainability.

Success stories and ongoing challenges

Success stories and ongoing challenges in the realm of environmental conservation and sustainability:

1. **Success Stories**:

    A. **Montreal Protocol**: One of the most celebrated success stories in environmental conservation, the Montreal Protocol is a global agreement aimed at phasing out ozone-depleting substances (ODSs). Since its adoption in 1987, the protocol has led to significant reductions in ODSs, resulting in gradual recovery of the ozone layer and preventing millions of cases of skin cancer and cataracts.

    B. **Renewable Energy Expansion**: The rapid expansion of renewable energy sources, such as solar, wind, and hydroelectric power, represents a major success story in the transition towards a low-carbon economy. Falling costs, technological advancements, and supportive policies have fueled the growth of renewable energy capacity worldwide, reducing greenhouse gas emissions and diversifying energy sources.

    C. **Conservation Successes**: Numerous conservation efforts have led to the recovery of endangered species and ecosystems around the world. For example, the revival of populations of species like the bald eagle, gray wolf, and southern white rhinoceros demonstrates the effectiveness of conservation strategies such as habitat protection, captive breeding, and reintroduction programs.

    D. **Environmental Policy Innovations**: Some countries have pioneered innovative environmental policies and initiatives that serve as models for others. For instance, Bhutan's commitment to maintaining 60% forest cover and prioritizing Gross National Happiness over Gross Domestic Product (GDP) has garnered international recognition as a holistic approach to sustainable development.

    E. **Community-Led Conservation**: Community-led conservation initiatives have proven effective in empowering local communities to protect and manage natural resources sustainably. Examples include community-managed marine protected areas, indigenous-led conservation projects, and grassroots movements advocating for land rights and environmental justice.

2. **Ongoing Challenges**:

   A. **Climate Change**: Climate change remains one of the most urgent and complex challenges facing humanity. Despite international agreements and commitments, global efforts to reduce greenhouse gas emissions have fallen short of the targets needed to limit global warming to safe levels. The impacts of climate change, including extreme weather events, sea-level rise, and disruptions to ecosystems, continue to escalate, posing significant risks to human societies and ecosystems worldwide.

   B. **Biodiversity Loss**: Biodiversity loss is accelerating at an alarming rate, driven by habitat destruction, overexploitation of natural resources, pollution, and climate change. Despite conservation efforts, species extinction rates are unprecedentedly high, and ecosystems are increasingly degraded and fragmented. Halting biodiversity loss and restoring ecosystems will require transformative changes in land-use practices, conservation strategies, and sustainable development policies.

   C. **Pollution and Waste**: Pollution and waste continue to pose significant threats to human health, ecosystems, and the environment. Air pollution, water contamination, plastic pollution, and soil degradation are pervasive problems that require coordinated action at local, national, and global levels. Addressing pollution and waste will necessitate regulatory measures, technological innovations, and changes in consumer behavior to minimize environmental impacts and promote sustainable resource management.

   D. **Inequality and Environmental Justice**: Environmental degradation disproportionately affects marginalized communities, exacerbating social inequalities and injustices. Vulnerable populations, including indigenous peoples, low-income communities, and people of color, often bear the brunt of environmental hazards and lack access to clean air, water, and land. Achieving environmental justice requires addressing systemic inequalities, amplifying marginalized voices, and ensuring equitable access to environmental resources and decision-making processes.

E. **Political Will and Global Cooperation**: Despite growing awareness of environmental issues, political will and global cooperation remain significant barriers to effective action. Conflicting interests, competing priorities, and geopolitical tensions hinder progress on key environmental challenges, including climate change, biodiversity conservation, and sustainable development. Overcoming these obstacles will require leadership, collaboration, and commitment from governments, businesses, civil society, and individuals worldwide.

while there have been notable successes in environmental conservation and sustainability, significant challenges persist. Addressing these challenges will require concerted efforts, innovative solutions, and collective action at all levels of society. By learning from past successes and confronting ongoing challenges with determination and resolve, we can build a more resilient, equitable, and sustainable future for generations to come.

Chapter Four

Local Initiatives

Community-based projects

Community-based projects are initiatives driven by local communities to address specific environmental, social, or economic challenges within their own neighborhoods or regions. These projects empower communities to take ownership of their development, leverage local knowledge and resources, and foster collaboration and solidarity among residents. Here's a detailed exploration of community-based projects and their key characteristics:

1. **Local Ownership and Participation:**
   - Community-based projects are initiated, implemented, and managed by local residents, organizations, or grassroots groups. By actively involving community members in decision-making

processes, these projects ensure that interventions are tailored to the specific needs, priorities, and aspirations of the community.
- o Local ownership fosters a sense of pride, agency, and responsibility among community members, motivating them to actively contribute their time, skills, and resources to project activities. This grassroots approach promotes sustainability and long-term impact by building capacity and resilience within the community.

2. **Holistic and Integrated Approach**:
   - o Community-based projects often adopt a holistic and integrated approach to address multifaceted challenges that intersect environmental, social, and economic dimensions. Rather than focusing narrowly on single issues, these projects aim to promote comprehensive solutions that address root causes and leverage synergies between different sectors.
   - o For example, a community-based project to improve access to clean water may incorporate components such as watershed protection, water conservation, sanitation education, and community health initiatives to address broader environmental and social determinants of health.

3. **Partnerships and Collaboration**:
   - o Community-based projects frequently involve collaboration and partnerships with diverse stakeholders, including local government agencies, non-profit organizations, academic institutions, businesses, and other community groups. These partnerships leverage complementary expertise, resources, and networks to maximize the impact and reach of project interventions.
   - o By fostering collaboration and coalition-building, community-based projects promote social cohesion, trust, and solidarity among stakeholders, leading to more effective and sustainable outcomes. Partnerships also enhance the scalability and replicability of successful approaches, enabling lessons learned to be shared and applied across different contexts.

4. **Empowerment and Capacity Building**:
    - Community-based projects prioritize empowerment and capacity building as core principles, aiming to strengthen the skills, knowledge, and leadership capabilities of community members. Through training workshops, educational programs, and hands-on learning experiences, participants gain the tools and confidence to actively engage in decision-making processes and drive positive change.
    - Empowerment extends beyond individual capacity building to include collective empowerment, whereby communities develop inclusive and democratic structures for decision-making, governance, and conflict resolution. This participatory approach fosters a sense of ownership, agency, and solidarity among community members, enabling them to overcome challenges collectively and build resilient, self-reliant communities.
5. **Sustainability and Long-Term Impact**:
    - Community-based projects prioritize sustainability and long-term impact by integrating principles of environmental stewardship, social equity, and economic resilience into project design and implementation. Rather than pursuing short-term gains or external dependencies, these projects seek to build local capacity, strengthen institutional structures, and foster enduring relationships within the community.
    - Sustainable outcomes may include improved livelihoods, enhanced environmental quality, strengthened social cohesion, and increased resilience to external shocks. By investing in human capital, social capital, and natural capital, community-based projects contribute to the long-term well-being and prosperity of communities, laying the foundation for a more sustainable and equitable future.

Examples of community-based projects include community gardens, renewable energy cooperatives, waste recycling programs, sustainable agriculture initiatives, eco-tourism ventures, and disaster preparedness and resilience projects.

Grassroots movements

Grassroots movements are collective efforts driven by ordinary people at the local level to advocate for social, political, environmental, or economic change. These movements emerge from the bottom-up, often in response to perceived injustices, grievances, or unmet needs within communities. Grassroots movements mobilize individuals, communities, and networks to raise awareness, build solidarity, and exert pressure on decision-makers to address systemic issues and pursue solutions that reflect the interests and aspirations of the people they represent. Here's a detailed exploration of grassroots movements and their key characteristics:

1. **Origin and Nature**:
    - Grassroots movements typically originate from within communities, arising organically in response to specific issues, concerns, or grievances. These movements are characterized by their decentralized structure, horizontal leadership, and inclusive participation, allowing individuals from diverse backgrounds to come together around shared goals and values.
    - Grassroots movements are often fueled by a sense of injustice, inequality, or disenfranchisement, motivating individuals to take collective action to challenge entrenched power structures, advocate for change, and reclaim agency over their lives and communities.
2. **Key Characteristics**:
    - **Local Focus**: Grassroots movements operate at the local level, addressing issues that directly affect communities and neighborhoods. While their impact may extend beyond local boundaries, grassroots movements are rooted in the lived experiences and realities of the people they represent.
    - **Participatory Democracy**: Grassroots movements prioritize principles of participatory democracy, empowering individuals to actively engage in decision-making processes, collective action, and advocacy efforts. Decision-making is often decentralized, consensus-based, and inclusive,

allowing for diverse voices and perspectives to be heard.
- **Horizontal Leadership**: Grassroots movements typically eschew hierarchical structures and top-down leadership models in favor of horizontal or distributed leadership. Leadership roles are fluid, rotating, and based on merit rather than formal authority, allowing for greater flexibility, adaptability, and inclusivity.
- **Coalition-Building**: Grassroots movements often seek to build alliances and coalitions with other social movements, community organizations, labor unions, and advocacy groups to amplify their impact and leverage collective power. By forging solidarity across diverse constituencies, grassroots movements strengthen their advocacy efforts and broaden their base of support.
- **Nonviolent Resistance**: Grassroots movements commonly employ nonviolent tactics and strategies to advance their goals and challenge injustice. Nonviolent resistance encompasses a wide range of actions, including protests, demonstrations, civil disobedience, boycotts, strikes, and grassroots organizing, aimed at disrupting the status quo and mobilizing public opinion.

3. **Impact and Outcomes**:
    - Grassroots movements have the potential to catalyze transformative change by raising awareness, shifting public discourse, and pressuring decision-makers to enact policy reforms or institutional changes. While their impact may vary depending on context, grassroots movements have historically played instrumental roles in advancing civil rights, environmental protection, labor rights, gender equality, and social justice.
    - Grassroots movements can achieve tangible outcomes, such as legislative victories, policy changes, corporate accountability, and shifts in public attitudes and behaviors. Even in cases where immediate policy gains are not realized, grassroots movements can contribute to broader societal shifts and cultural changes over time.

4. **Challenges and Opportunities**:
    - Grassroots movements face numerous challenges, including limited resources, political repression, co-optation by external actors, internal divisions, and burnout among activists. Overcoming these challenges requires resilience, strategic thinking, and sustained grassroots organizing efforts.
    - At the same time, grassroots movements present opportunities for social innovation, community empowerment, and democratic renewal. By harnessing the collective power of ordinary people, grassroots movements have the potential to drive systemic change, build resilient communities, and advance social justice and equity.

Examples of grassroots movements include the civil rights movement, environmental justice movement, labor movement, feminist movement, LGBTQ+ rights movement, indigenous rights movement, and anti-globalization movement. These movements demonstrate the power of grassroots organizing and collective action in challenging injustice, promoting social change, and building more inclusive and equitable societies.

Individuals making differences in their communities

Individuals have the power to make a profound difference in their communities by taking initiative, mobilizing resources, and inspiring others to create positive change. Here are some detailed examples of individuals who have made a difference in their communities:

1. **Wangari Maathai - The Green Belt Movement:**
   - Wangari Maathai, a Kenyan environmentalist, founded the Green Belt Movement in 1977 to address deforestation, soil erosion, and lack of access to firewood and clean water in rural communities.
   - Through the Green Belt Movement, Maathai mobilized women to plant trees, promote sustainable land management practices, and empower communities to protect their natural resources.
   - Over the years, the Green Belt Movement has planted millions of trees, restored degraded landscapes, empowered women, and improved livelihoods for thousands of people across Kenya and beyond.
2. **Trevor Noah - Education Initiatives:**
   - Trevor Noah, a South African comedian and television host, has used his platform to advocate for education and support disadvantaged youth in South Africa.
   - In partnership with the charity organization, the Trevor Noah Foundation, Noah has launched various initiatives to provide educational resources, scholarships, and mentorship programs for underprivileged children.
   - Through his philanthropic efforts, Noah has helped improve access to quality education, promote literacy, and empower young people to fulfill their potential and break the cycle of poverty.
3. **Jamila Raqib - Nonviolent Resistance Training:**
   - Jamila Raqib is the executive director of the Albert Einstein Institution, a non-profit organization dedicated to advancing the study and strategic use of nonviolent action in conflicts around the world.

- Raqib has worked tirelessly to train activists, organizers, and civil society groups in nonviolent resistance tactics and strategies to challenge authoritarian regimes, promote democracy, and defend human rights.
- Through her training programs and advocacy work, Raqib has empowered individuals and communities to effectively resist oppression, build movements for social change, and uphold democratic values in their societies.

4. **María Gunnoe - Environmental Activism:**
   - María Gunnoe, a community organizer and environmental activist from West Virginia, has led the fight against mountaintop removal mining, a destructive form of coal mining that devastates landscapes and pollutes waterways.
   - Gunnoe has worked tirelessly to raise awareness about the impacts of mountaintop removal mining on local communities, public health, and the environment.
   - Through her advocacy efforts, Gunnoe has successfully pressured government agencies and corporations to implement stricter regulations, enforce environmental laws, and protect vulnerable communities from the harmful effects of coal mining.

5. **Craig Kielburger - Child Rights Advocacy:**
   - Craig Kielburger, a Canadian activist and social entrepreneur, founded Free The Children (now known as WE Charity) at the age of 12 to combat child labor and promote child rights around the world.
   - Through Free The Children, Kielburger has mobilized young people to take action on issues such as access to education, clean water, healthcare, and economic empowerment for marginalized children and communities.
   - Kielburger's efforts have inspired millions of young people to get involved in social activism, volunteerism, and philanthropy, making a tangible impact on the lives of children and families in need.

Chapter Five

Technology and Innovation

Advances in renewable energy

Advances in renewable energy technologies have revolutionized the global energy landscape, offering sustainable alternatives to traditional fossil fuels and mitigating the impacts of climate change. Here's a detailed exploration of some key advances in renewable energy:

1. **Solar Photovoltaic (PV) Technology**:
    - Solar PV technology has seen significant advancements in efficiency, cost-effectiveness, and scalability over the past few decades. Advances in materials science, manufacturing processes, and system design have led to higher conversion

efficiencies, reduced production costs, and improved reliability of solar panels.
- Innovations such as thin-film solar cells, bifacial panels, and perovskite solar cells have expanded the range of applications for solar PV technology and increased its competitiveness with conventional energy sources.
- Integration of solar PV systems with energy storage technologies, smart grid technologies, and digital monitoring and control systems has enhanced grid reliability, flexibility, and resilience, enabling greater penetration of solar energy in the energy mix.

2. **Wind Energy**:
    - Wind energy has become increasingly competitive as a source of electricity generation, thanks to advancements in wind turbine design, manufacturing, and installation techniques. Modern wind turbines are larger, more efficient, and more reliable than their predecessors, capable of capturing more energy from the wind and operating in a wider range of wind conditions.
    - Offshore wind energy has emerged as a promising frontier in renewable energy development, offering vast untapped resources and higher energy yields compared to onshore wind farms. Technological innovations such as floating wind turbines, advanced blade designs, and dynamic control systems have made offshore wind projects more feasible and cost-effective.
    - Digitalization and predictive maintenance technologies have improved the performance and reliability of wind turbines, reducing downtime and operational costs and maximizing energy production over the lifetime of wind farms.

3. **Energy Storage**:
    - Advances in energy storage technologies have addressed one of the key challenges of renewable energy integration: intermittency and variability. Battery storage systems, in particular, have seen dramatic improvements in energy density, cycle life, and cost-effectiveness, making them increasingly viable for grid-scale energy storage applications.

- Lithium-ion batteries dominate the market for grid-scale energy storage, but other technologies such as flow batteries, pumped hydro storage, and thermal energy storage are also gaining traction. Research and development efforts are focused on improving the performance, safety, and sustainability of energy storage systems and reducing their environmental footprint.
- Integration of renewable energy sources with energy storage technologies enables greater flexibility, reliability, and resilience of power systems, facilitating the transition to a more decentralized, distributed, and renewable-based energy infrastructure.

4. **Bioenergy and Biofuels**:
   - Advances in bioenergy technologies, such as biomass conversion, biofuel production, and biogas generation, have expanded the range of feedstocks and conversion pathways for bioenergy production. Biogas plants, for example, can now utilize a wide variety of organic waste streams, including agricultural residues, food waste, and municipal solid waste, to produce renewable biogas for electricity generation, heat production, and transportation fuels.
   - Biofuel production processes have become more efficient and cost-effective, with innovations such as cellulosic ethanol production, algae-based biofuels, and synthetic biology approaches for biofuel synthesis. Sustainable bioenergy production practices, such as crop rotation, agroforestry, and waste-to-energy conversion, minimize environmental impacts and enhance the overall sustainability of bioenergy systems.

5. **Hydropower and Marine Energy**:
   - Hydropower remains the largest source of renewable electricity globally, providing reliable and dispatchable energy from flowing water. Advances in hydropower technology, such as advanced turbines, pumped storage systems, and run-of-river projects, have increased efficiency, flexibility, and environmental sustainability of hydropower installations.

- Marine energy technologies, including tidal energy, wave energy, and ocean thermal energy conversion (OTEC), hold immense potential for generating electricity from the ocean's renewable resources. Research and development efforts are focused on overcoming technical, environmental, and economic challenges to commercialize marine energy technologies and harness the vast energy potential of the world's oceans.

Sustainable practices in industry and agriculture

Sustainable practices in industry and agriculture are essential for mitigating environmental impacts, conserving natural resources, and promoting long-term economic viability. Here's a detailed exploration of some key sustainable practices in these sectors:

1. **Industry**:

    A. **Resource Efficiency and Conservation**:
    - Industry can adopt sustainable practices to improve resource efficiency and reduce waste generation. This includes implementing measures to optimize energy use, minimize water consumption, and reduce raw material inputs through recycling, reuse, and resource recovery.
    - Energy-efficient technologies, such as high-efficiency motors, lighting systems, and heating, ventilation, and air conditioning (HVAC) systems, can significantly reduce energy consumption and greenhouse gas emissions in industrial operations.
    - Water-saving technologies, such as water recycling and reuse systems, process optimization, and leak detection and repair programs, can help industries minimize water withdrawals and discharge volumes, conserve freshwater resources, and protect water quality.

    B. **Pollution Prevention and Control**:
    - Industries can implement pollution prevention measures to minimize emissions of air pollutants, water pollutants, and hazardous substances from their operations. This includes adopting cleaner production processes, installing pollution control technologies, and implementing best management practices for waste management and disposal.
    - Pollution control technologies, such as scrubbers, filters, and catalytic converters, can capture and treat air and water emissions to reduce environmental impacts and comply with regulatory requirements.

- Environmental management systems, such as ISO 14001 certification, provide a framework for industries to identify, assess, and mitigate environmental risks, improve environmental performance, and demonstrate commitment to sustainability.

C. **Circular Economy and Sustainable Supply Chains**:

- Embracing the principles of a circular economy, industries can minimize waste generation, maximize resource efficiency, and promote closed-loop material flows. This involves designing products for durability, repairability, and recyclability, and integrating recycled materials and renewable resources into production processes.
- Sustainable supply chain management practices, such as green procurement, supplier engagement, and product life cycle assessments, can help industries identify and mitigate environmental and social risks throughout the supply chain.
- Collaboration with suppliers, customers, and other stakeholders is essential for driving systemic changes in supply chains and promoting sustainability across industries and sectors.

2. **Agriculture**:

A. **Regenerative Agriculture**:

- Regenerative agriculture practices focus on restoring and enhancing soil health, biodiversity, and ecosystem resilience while improving agricultural productivity and profitability. These practices include crop rotation, cover cropping, conservation tillage, agroforestry, and integrated pest management.
- By building soil organic matter, enhancing nutrient cycling, and increasing soil water retention, regenerative agriculture improves soil fertility, reduces erosion, and enhances climate resilience, mitigating the impacts of climate change and extreme weather events.

B. **Agroecology and Sustainable Farming Practices**:
- Agroecological principles emphasize the integration of ecological concepts and practices into agricultural systems to promote biodiversity, ecosystem services, and food sovereignty. Agroecological farming practices include polyculture, agroforestry, organic farming, and diversified cropping systems.
- Sustainable farming practices aim to minimize environmental impacts, conserve natural resources, and promote ecological resilience while maintaining or increasing agricultural productivity and livelihoods. This includes reducing reliance on synthetic inputs, such as pesticides and fertilizers, and promoting biological pest control, soil conservation, and water management strategies.

C. **Climate-Smart Agriculture**:
- Climate-smart agriculture (CSA) approaches seek to address the triple challenge of food security, climate change adaptation, and mitigation in agricultural systems. CSA practices include climate-resilient crop varieties, water-efficient irrigation systems, agroforestry, and climate-smart livestock management.
- By enhancing the adaptive capacity of agricultural systems, improving resilience to climate variability, and reducing greenhouse gas emissions, climate-smart agriculture contributes to sustainable development goals and enhances the resilience of rural communities to climate change impacts.

D. **Precision Agriculture and Digital Technologies**:
- Precision agriculture technologies, such as remote sensing, geographic information systems (GIS), global positioning systems (GPS), and data analytics, enable farmers to optimize resource use, improve crop yields, and minimize environmental impacts through targeted interventions.
- Digital technologies, such as mobile applications, sensor networks, and farm management software,

provide farmers with real-time information and decision support tools to monitor crop health, manage inputs, and optimize farming operations, leading to more sustainable and efficient agricultural practices.

Adopting sustainable practices in industry and agriculture is critical for promoting environmental stewardship, enhancing resource efficiency, and building resilience to environmental challenges. By integrating sustainability principles into business models, production processes, and agricultural systems, industries and farmers can contribute to a more equitable, resilient, and sustainable future for people and the planet.

Tech solutions for environmental monitoring

Technological solutions for environmental monitoring and conservation have emerged as powerful tools for understanding, managing, and protecting natural ecosystems, biodiversity, and resources. These innovative technologies leverage advancements in sensors, data analytics, remote sensing, artificial intelligence (AI), and communication technologies to collect, analyze, and visualize environmental data, facilitate decision-making, and drive conservation efforts. Here's a detailed exploration of some key tech solutions for environmental monitoring and conservation:

1. **Remote Sensing and Earth Observation**:
    o Remote sensing technologies, such as satellites, drones (unmanned aerial vehicles), and aircraft-mounted sensors, provide valuable insights into changes in land cover, land use, vegetation health, and ecosystem dynamics over large spatial scales and temporal scales.
    o High-resolution satellite imagery, multispectral and hyperspectral sensors, and LiDAR (Light Detection and Ranging) technology enable scientists and conservationists to monitor deforestation, habitat loss, biodiversity hotspots, illegal logging, and land degradation in near real-time.
    o Earth observation platforms, such as NASA's Landsat and MODIS satellites, the European Space Agency's Sentinel satellites, and commercial satellite constellations, provide open-access data for a wide range of environmental applications, including monitoring climate change, assessing ecosystem health, and informing conservation planning and decision-making.

2. **Sensor Networks and Internet of Things (IoT)**:
   - Sensor networks and IoT technologies enable real-time monitoring of environmental parameters, such as air quality, water quality, soil moisture, temperature, and biodiversity, at local scales and high temporal resolutions.
   - Wireless sensor networks, environmental monitoring stations, and smart sensors deployed in natural reserves, wildlife habitats, and urban areas collect continuous data streams, detect anomalies, and trigger alerts in response to environmental disturbances or pollution events.
   - IoT-enabled devices, such as smart water meters, air quality monitors, and wildlife tracking tags, provide valuable insights into environmental conditions, habitat usage, species behavior, and ecological interactions, facilitating evidence-based decision-making and adaptive management strategies.
3. **Data Analytics and Machine Learning**:
   - Data analytics techniques, including statistical analysis, machine learning, and AI algorithms, process large volumes of environmental data, extract meaningful patterns and trends, and generate predictive models for environmental monitoring and forecasting.
   - Machine learning algorithms, such as neural networks, decision trees, and random forests, analyze satellite imagery, sensor data, and ecological datasets to identify habitat types, map land cover changes, predict species distributions, and assess ecosystem health and resilience.
   - Predictive modeling tools, such as species distribution models (SDMs) and ecological niche modeling (ENM), integrate environmental variables, species occurrence data, and habitat suitability criteria to predict species distributions, assess habitat connectivity, and prioritize conservation actions.
4. **Citizen Science and Crowdsourcing**:
   - Citizen science initiatives and crowdsourcing platforms engage volunteers, citizen scientists, and community members in collecting environmental data, monitoring biodiversity, and contributing to scientific research and conservation efforts.

- Mobile apps, online platforms, and social media networks enable citizen scientists to report wildlife sightings, record environmental observations, and participate in crowd-mapping projects, such as eBird, iNaturalist, and OpenStreetMap.
- Citizen science data complement traditional monitoring efforts, provide valuable insights into species distributions, phenology, and ecosystem dynamics, and foster public engagement, awareness, and stewardship of natural resources and ecosystems.

5. **Blockchain Technology for Conservation Finance and Supply Chain Transparency**:
   - Blockchain technology offers innovative solutions for addressing challenges related to conservation finance, sustainable supply chains, and wildlife trafficking by providing transparent, secure, and immutable records of transactions and data exchanges.
   - Blockchain-based platforms enable transparent tracking and verification of conservation investments, donations, and payments, ensuring that funds are allocated efficiently, and projects are implemented effectively.
   - Supply chain traceability solutions powered by blockchain technology allow stakeholders to track the origin, production, and distribution of sustainably sourced products, such as timber, seafood, and agricultural commodities, and verify compliance with environmental standards and certification schemes.

Chapter Six

Youth Activism

Role of young people in driving environmental change

The role of young people in driving environmental change is increasingly vital as they bring fresh perspectives, innovative ideas, and passionate activism to address pressing environmental challenges. Here's a detailed exploration of the significant contributions and roles young people play in driving environmental change:

1. **Youth Activism and Advocacy**:
    - Young people are at the forefront of environmental activism and advocacy, organizing protests, marches, and campaigns to raise awareness, mobilize support, and demand action on issues such as climate change, biodiversity loss, pollution, and environmental justice.

- Youth-led movements, such as Fridays for Future, Extinction Rebellion, and Youth Climate Strike, have galvanized millions of young people worldwide to take to the streets, demand climate action from policymakers, and hold governments and corporations accountable for their environmental policies and practices.
2. **Innovation and Entrepreneurship**:
    - Young innovators and entrepreneurs are driving technological advancements and developing sustainable solutions to address environmental challenges. From renewable energy startups to waste management initiatives, young entrepreneurs are leveraging technology, creativity, and social entrepreneurship to develop scalable and impactful solutions for a more sustainable future.
    - Incubators, accelerators, and innovation hubs dedicated to sustainable development provide platforms for young innovators to collaborate, receive mentorship, access funding, and scale up their ventures, fostering a culture of innovation and entrepreneurship in the environmental sector.
3. **Youth-Led Research and Science Communication**:
    - Young scientists, researchers, and science communicators play a crucial role in advancing scientific knowledge, conducting research on environmental issues, and communicating findings to the public, policymakers, and stakeholders.
    - Youth-led research initiatives, citizen science projects, and academic collaborations contribute to our understanding of environmental processes, ecosystem dynamics, and human impacts on the environment, informing evidence-based decision-making and policy formulation.
    - Social media platforms, blogs, podcasts, and science communication initiatives led by young people disseminate scientific information, raise awareness about environmental issues, and engage audiences in conversations about sustainability and conservation.
4. **Education and Youth Empowerment**:
    - Education plays a fundamental role in empowering young people with the knowledge, skills, and values to become environmental stewards and

changemakers. Environmental education programs, youth leadership initiatives, and youth empowerment platforms provide opportunities for young people to learn about sustainability, ecology, and environmental conservation and develop leadership skills and civic engagement.
- Youth organizations, clubs, and networks focused on environmental issues provide platforms for young people to collaborate, exchange ideas, and take collective action on environmental projects and campaigns at local, national, and global levels.

5. **Intergenerational Collaboration and Mentorship**:
   - Interactions and collaborations between young people and older generations are essential for driving environmental change and fostering intergenerational dialogue, mutual learning, and collective action.
   - Mentorship programs, internships, and cross-generational partnerships enable young people to benefit from the wisdom, experience, and guidance of seasoned environmentalists, activists, and professionals while providing fresh perspectives, energy, and enthusiasm to established organizations and movements.

6. **Policy Advocacy and Youth Representation**:
   - Young people are increasingly engaging in policy advocacy and political activism to influence decision-making processes, shape public policies, and advocate for legislation that promotes environmental sustainability, social justice, and equity.
   - Youth councils, youth parliaments, and youth-led organizations advocate for youth representation in decision-making bodies, government agencies, and international forums, ensuring that the voices and perspectives of young people are heard and integrated into policy discussions and decision-making processes.

Student-led movements and campaigns

Student-led movements and campaigns have historically played a significant role in driving social and political change, advocating for human rights, equality, and environmental sustainability. These movements mobilize young people to challenge injustices, demand accountability, and push for systemic reforms through collective action, advocacy, and activism. Here's a detailed exploration of student-led movements and campaigns:

1. **Historical Context**:
    - Student-led movements have a rich history dating back to the early 20th century, with examples such as the student protests against apartheid in South Africa, the Civil Rights Movement in the United States, and the student-led democracy movements in Eastern Europe.
    - In the 21st century, student-led movements have continued to shape political discourse and drive social change, with movements such as the Arab Spring, Occupy Wall Street, and Black Lives Matter gaining widespread attention and inspiring activism around the world.
2. **Environmental Advocacy and Climate Activism**:
    - In recent years, student-led movements focused on environmental advocacy and climate activism have gained momentum, with young people taking to the streets to demand urgent action on climate change and environmental degradation.
    - Initiatives such as Fridays for Future, started by Swedish activist Greta Thunberg, have mobilized millions of students worldwide to participate in climate strikes, protests, and campaigns to raise awareness about the climate crisis and demand policy action from governments and corporations.
3. **Student Divestment Campaigns**:
    - Student-led divestment campaigns have targeted universities, colleges, and other institutions to divest their endowments and investments from fossil fuel companies and other industries contributing to environmental degradation.

- The fossil fuel divestment movement, initiated by students at colleges and universities around the world, has pressured institutions to align their investments with their environmental and social values, leading to divestment commitments from hundreds of institutions globally.
4. **Campus Sustainability Initiatives**:
    - Students are driving sustainability initiatives on college and university campuses, advocating for renewable energy, zero waste, sustainable transportation, and environmentally friendly practices in campus operations and facilities management.
    - Student-led sustainability organizations, green clubs, and student government initiatives collaborate with campus administrators, faculty, and staff to implement sustainability policies, promote eco-friendly behaviors, and engage the campus community in environmental stewardship.
5. **Environmental Education and Awareness**:
    - Student-led campaigns focus on environmental education and awareness-raising to empower peers and communities to take action on environmental issues. These campaigns use creative strategies, such as workshops, seminars, film screenings, and social media campaigns, to educate and mobilize students around environmental sustainability.
    - Environmental education initiatives on college campuses foster a culture of sustainability, promote ecological literacy, and inspire future generations of environmental leaders and changemakers.
6. **Intersectional Advocacy and Social Justice**:
    - Student-led movements often intersect with broader social justice issues, advocating for intersectional approaches to address systemic inequalities and injustices. Environmental justice campaigns led by students prioritize marginalized communities disproportionately impacted by environmental hazards, pollution, and climate change.
    - Student activists recognize the interconnectedness of social, economic, and environmental issues and advocate for solutions that address root causes and

promote equity, justice, and solidarity across diverse communities and movements.

Youth voices and perspectives on the future of the planet

Youth voices and perspectives on the future of the planet are characterized by a sense of urgency, determination, and hope as young people confront the unprecedented environmental challenges and opportunities facing the world today. Here's a detailed exploration of youth perspectives on the future of the planet:

1. **Climate Crisis and Environmental Degradation:**
    - Young people are deeply concerned about the escalating climate crisis, environmental degradation, and loss of biodiversity, recognizing the existential threats these issues pose to human societies, ecosystems, and future generations.
    - Growing up in a world marked by extreme weather events, wildfires, rising sea levels, and ecological disruptions, young people feel a sense of urgency to address the root causes of environmental degradation and take bold action to mitigate climate change and protect the planet.
2. **Interconnectedness and Intersectionality:**
    - Youth perspectives on the future of the planet emphasize the interconnectedness of social, economic, and environmental issues and the importance of addressing systemic inequalities and injustices to achieve sustainability and equity.
    - Young people recognize the intersectionality of environmental challenges with other pressing issues, such as social justice, racial equity, gender equality, and economic development, and advocate for holistic approaches that prioritize human rights, dignity, and well-being alongside environmental conservation.
3. **Empowerment and Agency:**
    - Despite facing daunting environmental challenges, young people express a strong sense of agency, empowerment, and optimism about their ability to drive positive change and shape a more sustainable future.
    - Inspired by youth-led movements and campaigns, such as Fridays for Future and Extinction Rebellion, young activists are mobilizing peers, raising awareness, and demanding action from governments,

corporations, and other stakeholders to address environmental issues and uphold the rights of future generations.

4. **Innovation and Technological Solutions**:
   - Youth perspectives on the future of the planet emphasize the role of innovation, technology, and entrepreneurship in finding sustainable solutions to environmental challenges and driving positive change.
   - Young innovators and entrepreneurs are leveraging emerging technologies, such as renewable energy, artificial intelligence, and blockchain, to develop innovative solutions for clean energy, resource conservation, waste management, and climate resilience, demonstrating the transformative potential of technology in advancing sustainability goals.

5. **Intergenerational Collaboration and Solidarity**:
   - Youth perspectives underscore the importance of intergenerational collaboration, mentorship, and solidarity in addressing environmental issues and building a more sustainable world.
   - Young people seek guidance, mentorship, and support from older generations, while also challenging entrenched power structures, advocating for youth representation in decision-making processes, and pushing for meaningful participation in shaping policies and strategies for environmental conservation and sustainability.

6. **Education and Awareness**:
   - Youth voices emphasize the critical role of education, awareness-raising, and capacity-building in empowering young people to become environmental stewards, changemakers, and leaders in their communities.
   - Environmental education initiatives, youth-led campaigns, and peer-to-peer learning platforms provide opportunities for young people to deepen their understanding of environmental issues, develop critical thinking skills, and engage in hands-on conservation projects and advocacy efforts.

Chapter Seven

Conservation and Restoration

Efforts to protect ecosystems and endangered species

Efforts to protect ecosystems and endangered species are essential for safeguarding biodiversity, preserving ecosystem services, and ensuring the long-term health and resilience of natural ecosystems. These efforts encompass a wide range of conservation strategies, policies, and initiatives aimed at mitigating threats to ecosystems and species, restoring degraded habitats, and promoting sustainable management practices. Here's a detailed exploration of some key efforts to protect ecosystems and endangered species:

1. **Protected Areas and Habitat Conservation**:
    - Establishing and managing protected areas, such as national parks, wildlife reserves, marine protected areas, and nature reserves, is a cornerstone of ecosystem conservation and biodiversity protection.
    - Protected areas provide crucial habitats and refuges for endangered species, preserve biodiversity hotspots, and safeguard key ecological processes and ecosystem services, such as carbon sequestration, water filtration, and soil conservation.
    - Efforts to expand and strengthen protected area networks, enhance their connectivity, and improve their management effectiveness are essential for conserving ecosystems and preventing species extinctions.
2. **Species Conservation and Recovery Programs**:
    - Species conservation and recovery programs focus on identifying and prioritizing endangered species for conservation action, implementing targeted interventions to protect and restore their habitats, and

- monitoring population trends to assess conservation status and progress.
  - These programs employ a variety of conservation strategies, including captive breeding and reintroduction, habitat restoration, invasive species control, population monitoring, and community-based conservation initiatives.
  - Collaborative partnerships between government agencies, non-profit organizations, research institutions, and local communities are crucial for coordinating conservation efforts, pooling resources, and leveraging expertise to maximize conservation impact.
3. **Legal Protections and Policy Instruments**:
   - Legal protections and policy instruments, such as endangered species laws, wildlife trade regulations, habitat conservation plans, and environmental impact assessments, provide a framework for regulating human activities and preventing harm to ecosystems and species.
   - International agreements, such as the Convention on Biological Diversity (CBD), the Convention on International Trade in Endangered Species of Wild Fauna and Flora (CITES), and the Ramsar Convention on Wetlands, establish global frameworks for biodiversity conservation, species protection, and habitat preservation.
   - National and regional governments enact legislation, develop conservation policies, and establish regulatory mechanisms to enforce environmental laws, protect critical habitats, and mitigate threats to ecosystems and species from habitat destruction, pollution, poaching, and overexploitation.
4. **Community-Based Conservation and Indigenous Stewardship**:
   - Community-based conservation approaches engage local communities, indigenous peoples, and traditional resource users as stewards of natural resources, empowering them to participate in conservation decision-making, management, and benefit-sharing.
   - Indigenous peoples and local communities often possess traditional knowledge, cultural practices, and

sustainable land management techniques that contribute to the conservation and sustainable use of biodiversity.
- Collaborative conservation partnerships between conservation organizations and indigenous communities promote respect for indigenous rights, traditional governance systems, and local knowledge, fostering co-management of natural resources and building resilience to environmental change.

5. **Scientific Research and Monitoring**:
    - Scientific research and monitoring play a critical role in informing conservation strategies, assessing ecological health, and tracking changes in species populations, habitats, and ecosystems over time.
    - Ecological research, biodiversity surveys, and species monitoring programs provide essential data and insights into the status and trends of ecosystems and species, identifying conservation priorities, assessing threats, and evaluating the effectiveness of conservation interventions.
    - Advances in technology, such as remote sensing, GIS (Geographic Information Systems), camera traps, and DNA analysis, enhance our ability to monitor ecosystems, track wildlife movements, and detect changes in biodiversity at local and landscape scales, supporting evidence-based conservation decision-making.

6. **Public Awareness and Education**:
    - Public awareness and education campaigns raise awareness about the value of biodiversity, the importance of ecosystems, and the threats facing endangered species, inspiring public support for conservation efforts and fostering a sense of stewardship for the natural world.
    - Environmental education programs, outreach initiatives, and community engagement activities promote ecological literacy, empower individuals to take action for conservation, and cultivate a culture of environmental responsibility and sustainability.

Restoration projects

Restoration projects for forests, wetlands, and other habitats are essential for reversing habitat loss, enhancing biodiversity, and promoting ecosystem resilience in the face of environmental degradation and climate change. These projects involve a variety of interventions aimed at restoring degraded ecosystems, rehabilitating habitats, and revitalizing ecosystem functions and services. Here's a detailed exploration of restoration projects for forests, wetlands, and other habitats:

1. **Forest Restoration**:
    - Forest restoration projects focus on restoring degraded forests, recovering forest ecosystems, and enhancing the ecological integrity and resilience of forest landscapes.
    - Afforestation and reforestation efforts involve planting native tree species, restoring forest cover on deforested or degraded lands, and reconnecting fragmented forest habitats to improve biodiversity, soil stabilization, and carbon sequestration.
    - Sustainable forest management practices, such as selective logging, agroforestry, and natural regeneration, promote ecosystem health, timber production, and livelihoods while conserving biodiversity and protecting forest ecosystems.
2. **Wetland Restoration**:
    - Wetland restoration projects aim to restore and rehabilitate degraded wetland habitats, including marshes, swamps, bogs, and mangrove forests, to improve water quality, biodiversity, and ecosystem services.

- Wetland restoration techniques include restoring hydrological connectivity, reintroducing native vegetation, controlling invasive species, and implementing water management strategies to restore natural water flow regimes and enhance wetland functions and values.
- Restoring wetlands helps mitigate flood risks, improve water retention and filtration, support migratory bird populations, and provide critical habitat for aquatic species, contributing to climate adaptation and resilience.

3. **Riparian Restoration**:
   - Riparian restoration projects focus on restoring riparian zones, the transitional areas between terrestrial and aquatic ecosystems, along rivers, streams, and waterways.
   - Riparian restoration interventions include planting native vegetation along riverbanks, stabilizing stream channels, restoring natural floodplain connectivity, and implementing erosion control measures to improve water quality, aquatic habitat, and ecosystem connectivity.
   - Restoring riparian habitats enhances biodiversity, supports aquatic ecosystems, reduces sedimentation and nutrient runoff, and provides important ecological services, such as flood regulation, water purification, and groundwater recharge.

4. **Grassland Restoration**:
   - Grassland restoration projects aim to restore native grassland habitats, savannas, and prairies, which are among the most threatened and degraded ecosystems globally.
   - Grassland restoration techniques include reseeding with native grass species, managing grazing regimes, controlling invasive species, and restoring fire regimes to promote biodiversity, soil health, and ecosystem resilience.
   - Restoring grassland habitats supports grassland-dependent species, such as grassland birds, pollinators, and herbivores, enhances carbon sequestration and soil carbon storage, and provides ecosystem services, such as soil erosion control and water infiltration.

5. **Coral Reef Restoration:**
    - Coral reef restoration projects aim to restore and rehabilitate degraded coral reef ecosystems, which are facing unprecedented threats from climate change, ocean acidification, pollution, and overfishing.
    - Coral reef restoration techniques include coral propagation, transplantation, and larval rearing, as well as habitat restoration measures, such as artificial reef structures, coral nurseries, and marine protected areas.
    - Restoring coral reef ecosystems enhances biodiversity, supports fisheries, protects coastal communities from storm surges and erosion, and promotes ecotourism and sustainable livelihoods, contributing to coastal resilience and sustainable development.
6. **Community-Based Restoration:**
    - Community-based restoration projects engage local communities, indigenous peoples, and stakeholders as partners in conservation and restoration efforts, fostering stewardship, ownership, and sustainability.
    - These projects involve participatory planning, capacity-building, and collaborative decision-making processes that empower communities to identify conservation priorities, design restoration interventions, and manage natural resources sustainably.
    - Community-based restoration initiatives promote social equity, cultural resilience, and environmental justice by recognizing and respecting local knowledge, traditional practices, and customary rights to land and resources.

Importance of preserving natural spaces

Preserving natural spaces is of paramount importance for safeguarding biodiversity, promoting ecosystem services, and fostering human well-being. Natural spaces, including forests, wetlands, grasslands, and marine habitats, provide a wide range of ecological, social, economic, and cultural benefits that are essential for sustaining life on Earth. Here's a detailed exploration of the importance of preserving natural spaces:

1. **Biodiversity Conservation:**
   - Natural spaces harbor diverse ecosystems, species, and genetic diversity, playing a critical role in conserving biodiversity and preserving the web of life.
   - Protected areas, such as national parks, wildlife reserves, and nature reserves, provide habitats and refuges for endangered species, endemic species, and migratory wildlife, safeguarding biodiversity hotspots and preserving ecological processes and evolutionary processes.
2. **Ecosystem Services:**
   - Natural spaces provide essential ecosystem services that support human well-being, including provisioning services (e.g., food, water, timber), regulating services (e.g., climate regulation, water purification), supporting services (e.g., soil formation, nutrient cycling), and cultural services (e.g., recreation, spiritual and cultural values).
   - Wetlands, forests, and coastal habitats contribute to flood regulation, erosion control, and water purification, reducing the risks of natural disasters and providing valuable ecosystem services to local communities and societies.
3. **Climate Regulation and Carbon Sequestration:**
   - Natural spaces play a crucial role in regulating the Earth's climate by sequestering carbon dioxide from the atmosphere, mitigating climate change, and buffering against its impacts.
   - Forests, mangroves, and peatlands are particularly effective carbon sinks, storing large amounts of carbon in biomass, soils, and organic matter, and

helping to stabilize global carbon cycles and climate patterns.
4. **Water Resources and Watershed Protection**:
   - Natural spaces play a vital role in maintaining water quality, regulating hydrological cycles, and providing freshwater resources for human consumption, agriculture, industry, and ecosystem functions.
   - Forests, wetlands, and riparian zones act as natural water filters, trapping sediments, nutrients, and pollutants, and reducing the risks of water pollution and contamination downstream.
5. **Recreation and Ecotourism**:
   - Natural spaces provide opportunities for outdoor recreation, ecotourism, and nature-based tourism, contributing to local economies, livelihoods, and cultural heritage.
   - National parks, wildlife reserves, and protected areas attract visitors from around the world, generating revenue, creating jobs, and supporting small businesses in nearby communities, while also promoting environmental awareness and appreciation.
6. **Cultural and Spiritual Values**:
   - Natural spaces hold cultural, spiritual, and aesthetic significance for indigenous peoples, local communities, and societies, providing inspiration, solace, and a sense of connection to the natural world.
   - Sacred sites, traditional landscapes, and culturally significant areas preserve indigenous knowledge, cultural heritage, and traditional practices, fostering cultural resilience and identity.
7. **Education and Research**:
   - Natural spaces serve as living laboratories and outdoor classrooms for environmental education, scientific research, and experiential learning, fostering curiosity, discovery, and a deeper understanding of the natural world.
   - Field research, ecological monitoring, and biodiversity surveys conducted in natural spaces contribute to scientific knowledge, inform conservation strategies, and provide valuable insights into ecosystem dynamics and resilience.

Chapter Eight

Sustainable Living

Tips for reducing personal environmental impact

Reducing personal environmental impact is essential for mitigating climate change, conserving natural resources, and promoting sustainable living practices. Individuals can make a significant difference by adopting eco-friendly habits and making conscious choices in their daily lives. Here are some detailed tips for reducing personal environmental impact:

1. **Reduce Energy Consumption**:
    - Use energy-efficient appliances and lighting fixtures to reduce electricity consumption.

- Turn off lights, electronics, and appliances when not in use, and unplug chargers to prevent standby power consumption.
- Set thermostats to energy-saving temperatures and use programmable thermostats to regulate heating and cooling systems efficiently.
- Maximize natural lighting and ventilation by opening windows and using daylight whenever possible.

2. **Conserve Water**:
   - Fix leaks in faucets, toilets, and pipes to prevent water wastage.
   - Install water-saving devices, such as low-flow showerheads and faucet aerators, to reduce water usage.
   - Take shorter showers, turn off the tap while brushing teeth or washing dishes, and collect rainwater for outdoor watering.
   - Use water-efficient appliances, such as washing machines and dishwashers, and run full loads to minimize water consumption.

3. **Reduce, Reuse, Recycle**:
   - Reduce waste by opting for reusable products, such as water bottles, shopping bags, and containers, instead of single-use items.
   - Reuse and repurpose items whenever possible, such as jars, containers, and clothing, to extend their lifespan and reduce the need for new purchases.
   - Recycle paper, glass, plastic, and metal materials according to local recycling guidelines, and compost organic waste to divert it from landfills.

4. **Choose Sustainable Transportation**:
   - Use public transportation, carpooling, biking, or walking for daily commutes and errands to reduce carbon emissions and traffic congestion.
   - Choose fuel-efficient vehicles or consider alternative transportation options, such as electric or hybrid cars, for longer trips.
   - Plan and consolidate trips to minimize mileage and optimize fuel efficiency, and maintain vehicles regularly to ensure optimal performance.

5. **Eat Sustainably**:
   - Choose locally grown, seasonal, and organic foods whenever possible to support local farmers and

reduce the environmental footprint of food production and transportation.
- Reduce meat consumption and incorporate more plant-based foods into your diet to lower greenhouse gas emissions associated with livestock farming.
- Minimize food waste by planning meals, storing food properly, and composting organic scraps, and support food recovery programs to redistribute surplus food to those in need.

6. **Support Eco-Friendly Products and Practices**:
   - Purchase eco-friendly, fair trade, and sustainably sourced products that prioritize environmental and social responsibility.
   - Avoid products with excessive packaging or harmful chemicals, and opt for environmentally friendly alternatives, such as biodegradable cleaners and reusable household items.
   - Choose products made from renewable materials, such as bamboo, hemp, or recycled materials, and support companies that prioritize eco-friendly manufacturing and supply chain practices.

7. **Educate and Advocate**:
   - Stay informed about environmental issues, policies, and initiatives, and educate others about the importance of sustainable living and environmental stewardship.
   - Advocate for environmentally responsible policies and practices at the local, national, and global levels by supporting environmental organizations, participating in community events, and engaging with policymakers.
   - Lead by example and inspire others to adopt eco-friendly habits and behaviors through your actions, choices, and advocacy efforts.

Eco-friendly habits and practices

Adopting eco-friendly habits and practices is crucial for reducing our environmental footprint, conserving natural resources, and promoting sustainability in our daily lives. By making conscious choices and incorporating environmentally friendly behaviors into our routines, we can contribute to a healthier planet and a more sustainable future. Here are some detailed eco-friendly habits and practices:

1. **Reduce Energy Consumption**:
   - Use energy-efficient appliances and lighting: Replace incandescent bulbs with LED or CFL lights, and choose Energy Star-rated appliances whenever possible.
   - Turn off lights and electronics when not in use, and unplug chargers and devices to prevent standby power consumption.
   - Use natural lighting and ventilation: Open curtains during the day to let in sunlight, and use fans instead of air conditioning whenever possible.
2. **Conserve Water**:
   - Fix leaks in faucets, toilets, and pipes promptly to prevent water wastage.

- Install water-saving devices, such as low-flow showerheads and faucet aerators, to reduce water usage.
- Take shorter showers, turn off the tap while brushing teeth or washing dishes, and collect rainwater for outdoor watering.

3. **Reduce, Reuse, Recycle**:
    - Reduce waste by opting for reusable products, such as water bottles, shopping bags, and containers, instead of single-use items.
    - Reuse and repurpose items whenever possible: Use glass jars for storage, repair and refurbish old furniture and electronics, and donate or sell items you no longer need.
    - Recycle paper, glass, plastic, and metal materials according to local recycling guidelines, and compost organic waste to create nutrient-rich soil.

4. **Choose Sustainable Transportation**:
    - Use public transportation, carpooling, biking, or walking for daily commutes and errands to reduce carbon emissions and traffic congestion.
    - Choose fuel-efficient vehicles or consider alternative transportation options, such as electric or hybrid cars, for longer trips.
    - Plan and consolidate trips to minimize mileage and optimize fuel efficiency, and maintain vehicles regularly to ensure optimal performance.

5. **Eat Sustainably**:
    - Choose locally grown, seasonal, and organic foods whenever possible to support local farmers and reduce the environmental footprint of food production and transportation.
    - Reduce meat consumption and incorporate more plant-based foods into your diet to lower greenhouse gas emissions associated with livestock farming.
    - Minimize food waste by planning meals, storing food properly, and composting organic scraps, and support food recovery programs to redistribute surplus food to those in need.

6. **Conserve Resources**:
    - Use cloth napkins and towels instead of disposable paper products, and opt for cloth diapers instead of disposable ones for babies.

- Choose products made from renewable materials, such as bamboo, hemp, or recycled materials, and avoid products with excessive packaging or harmful chemicals.
- Repair and maintain items instead of replacing them: Mend clothing, fix appliances, and refurbish furniture to extend their lifespan and reduce waste.

7. **Support Eco-Friendly Practices**:
    - Support businesses and companies that prioritize environmental sustainability and social responsibility in their operations and supply chains.
    - Patronize farmers markets, local shops, and businesses that offer eco-friendly products and services, and choose fair trade and ethical brands whenever possible.
    - Educate yourself and others about environmental issues, policies, and initiatives, and advocate for environmentally responsible practices and policies in your community and beyond.

Green consumerism and ethical choices

Green consumerism and ethical choices involve making purchasing decisions that prioritize environmental sustainability, social responsibility, and ethical considerations. By choosing products and services that have minimal environmental impact, support fair labor practices, and promote social equity, individuals can contribute to positive change and encourage businesses to adopt more sustainable and ethical practices. Here's a detailed exploration of green consumerism and ethical choices:

1. **Environmental Sustainability**:
    o Green consumerism emphasizes the importance of selecting products and services that minimize environmental impact throughout their lifecycle, from production and distribution to use and disposal.
    o Environmentally friendly products are often made from renewable or recycled materials, use energy-efficient manufacturing processes, reduce waste and pollution, and are designed for longevity and recyclability.
    o Examples of environmentally sustainable products include organic and locally sourced food, eco-friendly cleaning supplies, energy-efficient appliances, and renewable energy options, such as solar panels and wind turbines.
2. **Fair Labor Practices**:
    o Ethical consumerism involves choosing products and brands that uphold fair labor practices, treat workers ethically, and ensure safe and dignified working conditions throughout the supply chain.
    o Ethical brands often adhere to labor standards established by international organizations, such as the International Labour Organization (ILO), and

may be certified by independent organizations, such as Fair Trade International or the Fair Wear Foundation.
- By supporting companies that prioritize fair labor practices and worker rights, consumers can help promote social justice, reduce exploitation, and improve livelihoods for workers in various industries, including agriculture, manufacturing, and retail.

3. **Social Responsibility**:
   - Ethical consumerism extends beyond environmental and labor considerations to encompass broader social responsibilities, such as supporting diversity, inclusion, and community development.
   - Socially responsible companies may engage in philanthropy, donate a portion of their profits to charitable causes, or support community development initiatives, such as education, healthcare, and poverty alleviation programs.
   - Consumers can choose to support socially responsible brands and businesses that align with their values and contribute positively to society, whether through their business practices, corporate social responsibility (CSR) initiatives, or community engagement efforts.

4. **Transparency and Accountability**:
   - Transparent and accountable companies provide clear and accurate information about their products, supply chains, and business practices, allowing consumers to make informed choices and hold companies accountable for their actions.
   - Ethical brands often disclose information about their sourcing practices, environmental impact, labor conditions, and social initiatives, enabling consumers to assess the ethical and sustainability credentials of products and brands.
   - Consumers can support transparent and accountable companies by seeking out information, asking questions, and demanding transparency and accountability from brands and businesses in their purchasing decisions.

5. **Educating and Empowering Consumers**:
   - Education plays a crucial role in empowering consumers to make ethical and sustainable choices by

raising awareness about environmental and social issues, providing information about sustainable alternatives, and promoting responsible consumption habits.
- Consumer advocacy organizations, non-profit groups, and independent certification schemes, such as B Corp certification and the Rainforest Alliance, provide resources, tools, and guidance to help consumers make ethical and sustainable purchasing decisions.
- By educating themselves about the impacts of their consumption choices and advocating for ethical and sustainable practices, consumers can drive positive change, influence market trends, and hold businesses accountable for their social and environmental performance.

Chapter Nine

The Future of Earth Day

Reflections on the evolution of Earth Day

Earth Day, celebrated annually on April 22nd, marks a global observance of environmental awareness and action. Since its inception in 1970, Earth Day has evolved significantly, reflecting changes in environmental consciousness, policy priorities, and global challenges. Here's a detailed reflection on the evolution of Earth Day since its inception:

1. **Origins and Early Years (1970s)**:
    - Earth Day was founded in 1970 by Senator Gaylord Nelson of Wisconsin, who envisioned a national day of environmental education and activism to raise awareness about pressing environmental issues.
    - The first Earth Day mobilized millions of Americans in rallies, demonstrations, and educational events across the country, sparking a grassroots movement that led to the creation of the U.S. Environmental Protection Agency (EPA) and the passage of landmark environmental legislation, including the Clean Air Act, Clean Water Act, and Endangered Species Act.
    - Earth Day in the 1970s reflected growing concerns about air and water pollution, deforestation, and habitat destruction, as well as the emergence of the modern environmental movement and the recognition of environmental issues as urgent and interconnected challenges requiring collective action.
2. **Global Expansion and Environmental Awareness (1980s-1990s)**:
    - Earth Day expanded internationally in the 1980s and 1990s, with countries around the world adopting

- April 22nd as a day to celebrate environmental awareness and promote sustainability.
  - Earth Day celebrations focused on a broader range of environmental issues, including biodiversity conservation, climate change, ozone depletion, and sustainable development, reflecting the growing recognition of the interconnectedness of environmental challenges and the need for global cooperation.
  - Earth Day in the 1980s and 1990s saw increased emphasis on environmental education, public outreach, and corporate engagement, with businesses, schools, and communities organizing events and initiatives to promote environmental stewardship and sustainability.
3. **Renewed Focus on Climate Change and Global Challenges (2000s-2010s)**:
   - In the 21st century, Earth Day evolved to address emerging environmental threats, such as climate change, ocean pollution, deforestation, and biodiversity loss, which became increasingly urgent and interconnected global challenges.
   - Earth Day campaigns and initiatives focused on raising awareness about climate change impacts, promoting renewable energy and energy efficiency, advocating for policy action, and mobilizing public support for international agreements, such as the Paris Agreement on climate change.
   - Earth Day in the 2000s and 2010s saw the rise of digital activism and online mobilization, with social media platforms and digital technologies enabling broader participation, engagement, and collaboration in environmental advocacy and activism.
4. **Intersectionality and Environmental Justice (2020s and Beyond)**:
   - Earth Day in the 2020s has brought increased attention to intersectional issues of environmental justice, equity, and social inequality, highlighting the disproportionate impacts of environmental degradation and climate change on marginalized communities, Indigenous peoples, and vulnerable populations.

- Earth Day campaigns and initiatives now emphasize the importance of addressing systemic inequalities, promoting environmental justice, and centering the voices and experiences of frontline communities in environmental advocacy and decision-making processes.
- Earth Day in the 2020s and beyond calls for a holistic approach to sustainability that integrates environmental, social, and economic considerations, fosters collaboration across sectors and stakeholders, and advances solutions that prioritize equity, resilience, and justice for all.

Vision for the future of environmental activism

The vision for the future of environmental activism is one of bold, inclusive, and transformative action that addresses the interconnected challenges of environmental degradation, social injustice, and climate change while promoting sustainability, resilience, and equity for all. Here's a detailed exploration of the vision for the future of environmental activism:

1. **Intersectionality and Environmental Justice**:
    - Environmental activism of the future embraces intersectionality and recognizes the interconnectedness of environmental issues with social justice, racial equity, and economic inequality.
    - The vision for environmental activism prioritizes the voices and experiences of marginalized communities, Indigenous peoples, and frontline populations who bear the brunt of environmental degradation and climate impacts.
    - Environmental justice principles guide activism efforts, ensuring that solutions address systemic inequalities, promote equitable access to resources, and empower communities to participate in decision-making processes that affect their lives and environments.

2. **Holistic Approach to Sustainability:**
   - Environmental activism of the future adopts a holistic approach to sustainability that integrates environmental, social, and economic dimensions, fostering a more balanced and equitable relationship between people and the planet.
   - Activists advocate for policies and practices that promote regenerative agriculture, circular economies, sustainable consumption and production, and renewable energy sources, while also addressing issues of poverty, inequality, and social exclusion.
   - The vision for environmental activism emphasizes the importance of valuing nature, respecting indigenous knowledge, and restoring ecosystems to support biodiversity, ecosystem services, and planetary health.

3. **Climate Action and Resilience:**
   - Environmental activism of the future places climate action at the forefront, advocating for ambitious mitigation and adaptation measures to address the climate crisis and build resilience to its impacts.
   - Activists call for rapid decarbonization of the economy, investments in renewable energy and clean technologies, and transition to sustainable transportation and land use practices to limit global warming and reduce greenhouse gas emissions.
   - The vision for environmental activism prioritizes climate justice, ensuring that climate policies and solutions benefit the most vulnerable communities and uphold principles of equity, fairness, and solidarity.

4. **Youth Empowerment and Intergenerational Collaboration:**
   - Environmental activism of the future empowers youth as agents of change and engages intergenerational collaboration to amplify voices, share knowledge, and mobilize collective action across generations.
   - Youth-led movements and campaigns play a central role in driving environmental activism, advocating for bold policy reforms, mobilizing public support, and holding governments and corporations accountable for their environmental and social responsibilities.

- Intergenerational dialogue, mentorship, and solidarity foster collaboration, learning, and innovation, ensuring that diverse perspectives and experiences are represented and valued in environmental decision-making and action.
5. **Global Cooperation and Solidarity**:
    - Environmental activism of the future embraces global cooperation, solidarity, and partnerships to address transboundary environmental challenges and promote collective solutions at local, national, and international levels.
    - Activists advocate for multilateral agreements, such as the Paris Agreement and the Convention on Biological Diversity, that set ambitious goals, provide mechanisms for accountability, and mobilize resources for climate action, biodiversity conservation, and sustainable development.
    - The vision for environmental activism emphasizes the importance of solidarity across borders, cultures, and communities, recognizing that environmental issues are global in nature and require collective action and shared responsibility to address effectively.

Call to action for readers

Getting involved and making a difference in environmental conservation and sustainability efforts is crucial for addressing the urgent challenges facing our planet. Here's a detailed call to action for readers to get involved and make a difference:

1. **Educate Yourself**: Start by educating yourself about environmental issues, climate change, biodiversity loss, and sustainability solutions. Stay informed through reputable sources, books, documentaries, and online resources. Understanding the issues is the first step towards taking meaningful action.
2. **Raise Awareness**: Share what you learn with friends, family, and your community. Raise awareness about environmental issues through social media, conversations, and events. Use your voice to advocate for positive change and inspire others to join the movement.
3. **Take Action Locally**: Get involved in local environmental initiatives, clean-up events, tree planting activities, and community gardens. Support local conservation organizations, volunteer at environmental nonprofits, and participate in local advocacy campaigns. Making a difference starts in your own backyard.
4. **Reduce Your Environmental Footprint**: Make sustainable choices in your daily life to reduce your environmental impact. Use energy-efficient appliances, reduce water usage, minimize waste, recycle and compost, choose eco-friendly products, and opt for sustainable transportation options. Every small change counts towards building a more sustainable future.
5. **Support Sustainable Businesses**: Support businesses that prioritize environmental sustainability, ethical sourcing, and social responsibility. Choose products and services from companies committed to reducing their carbon footprint, supporting fair labor practices, and minimizing environmental harm.
6. **Advocate for Policy Change**: Advocate for stronger environmental policies and regulations at the local, national, and global levels. Write to your elected representatives, participate in advocacy campaigns, and support

environmental organizations working to influence policy decisions and promote sustainable solutions.
7. **Invest in Sustainability**: Consider investing in companies and projects that promote sustainability, renewable energy, and environmental conservation. Support green technologies, sustainable agriculture, and clean energy initiatives that contribute to a more sustainable and resilient future.
8. **Engage in Collective Action**: Join environmental organizations, community groups, and grassroots movements working towards environmental justice, climate action, and biodiversity conservation. Participate in rallies, marches, and protests to demand bold action on environmental issues and hold governments and corporations accountable.
9. **Lifelong Learning and Improvement**: Commit to lifelong learning and continuous improvement in your environmental stewardship efforts. Stay engaged, open to new ideas, and willing to adapt your behaviors and habits in response to new information and evolving challenges.
10. **Inspire Others**: Be a role model for environmental stewardship and inspire others to take action. Lead by example, share your experiences and successes, and encourage others to join you in making a positive difference for the planet.

Hope for the planet and its inhabitants

Hope for the planet and its inhabitants lies in our collective ability to come together, take action, and create positive change for a sustainable and thriving future. Here's a detailed exploration of the hope we hold for the planet and its inhabitants:

1. **Resilience of Nature**:
   o Despite the environmental challenges we face, nature has shown remarkable resilience and capacity for regeneration. Ecosystems have the ability to rebound and adapt to changing conditions, given the opportunity and support.
2. **Innovative Solutions**:
   o Human ingenuity and innovation offer hope for addressing environmental issues through sustainable technologies, renewable energy solutions, green infrastructure, and conservation practices. Advances in science, engineering, and technology provide promising avenues for mitigating climate change, protecting biodiversity, and promoting sustainability.
3. **Youth Engagement and Activism**:
   o The growing youth movement for climate action and environmental justice represents a source of hope and inspiration. Young people around the world are leading grassroots movements, advocating for policy change, and driving collective action to address environmental challenges and promote a more sustainable future.
4. **Global Cooperation and Solidarity**:
   o The increasing recognition of environmental issues as global challenges requiring collective action and cooperation offers hope for meaningful progress. International agreements, partnerships, and collaborative initiatives provide platforms for countries, communities, and stakeholders to work

together towards common goals of environmental conservation and sustainability.

5. **Community Resilience and Empowerment:**
   - Local communities play a crucial role in building resilience, fostering sustainability, and promoting environmental stewardship. Community-based initiatives, grassroots movements, and participatory approaches empower individuals and communities to take ownership of their environmental future and drive positive change from the ground up.

6. **Regenerative Agriculture and Restoration:**
   - The adoption of regenerative agriculture practices, ecosystem restoration projects, and nature-based solutions offers hope for restoring degraded landscapes, enhancing biodiversity, and sequestering carbon. By working with nature rather than against it, we can regenerate ecosystems, improve soil health, and enhance ecosystem services for the benefit of all.

7. **Cultural Connections and Indigenous Wisdom:**
   - Indigenous peoples and local communities possess traditional knowledge, cultural practices, and spiritual connections to the land that offer valuable insights into sustainable living and environmental stewardship. By respecting and honoring indigenous wisdom, we can learn from their experiences and traditions to inform more holistic and harmonious approaches to conservation and sustainability.

8. **Shift in Values and Consciousness:**
   - There is a growing shift in values and consciousness towards greater environmental awareness, empathy, and responsibility. As more people recognize the intrinsic value of nature, the interconnectedness of all life, and the importance of living in harmony with the Earth, we can collectively cultivate a culture of care, compassion, and sustainability.

9. **Hope in Action:**
   - Ultimately, hope for the planet and its inhabitants lies in action. By taking meaningful steps towards sustainability in our daily lives, communities, and societies, we can contribute to positive change and create a brighter, more resilient future for ourselves and future generations. Each action, no matter how small, has the potential to make a difference and

inspire others to join in the effort to protect and preserve our precious planet.

www.ingramcontent.com/pod-product-compliance
Lightning Source LLC
Chambersburg PA
CBHW070351230526
45471CB00006B/2515